Bibliographic information published by the German National Library:

The German National Library lists this publication in the National Bibliography; detailed bibliographic data are available on the Internet at http://dnb.dnb.de .

Imprint:

Copyright © 2017 GRIN Verlag, Open Publishing GmbH
Print and binding: Books on Demand GmbH, Norderstedt Germany
ISBN: 9783668545618

This book at GRIN:

http://www.grin.com/en/e-book/376497/fast-active-queue-management-stability-transmission-control-protocol-fast

Christo Ananth

Fast Active Queue Management Stability Transmission Control Protocol (FAST TCP)

A Project Report

GRIN Publishing

GRIN - Your knowledge has value

Since its foundation in 1998, GRIN has specialized in publishing academic texts by students, college teachers and other academics as e-book and printed book. The website www.grin.com is an ideal platform for presenting term papers, final papers, scientific essays, dissertations and specialist books.

Visit us on the internet:

http://www.grin.com/

http://www.facebook.com/grincom

http://www.twitter.com/grin_com

FAST ACTIVE QUEUE MANAGEMENT STABILITY

TRANSMISSION CONTROL PROTOCOL

(FAST TCP)

A PROJECT REPORT

Submitted by

CHRISTO ANANTH

ELECTRONICS & COMMUNICATION ENGINEERING

NOORUL ISLAM COLLEGE OF ENGINEERING

ANNA UNIVERSITY: CHENNAI 600 025

Contents

I. INTRODUCTION

Congestion control is a distributed algorithm to share network resources among competing users. It is important in situations where the availability of resources and the set of competing users vary over time unpredictably, yet efficient sharing is desired. These constraints, unpredictable supply and demand and efficient operation, necessarily lead to feedback control as the preferred approach, where traffic sources dynamically adapt their rates to congestion in their paths. On the Internet, this is performed by the Transmission Control Protocol (TCP) in source and destination computers involved in data transfers. The congestion control algorithm in the current TCP, which we refer to as Reno, was developed in 1988 and has gone through several enhancements since. It has performed remarkably well and is generally believed to have prevented severe congestion as the Internet scaled up by six orders of magnitude in size, speed, load, and connectivity, if is also well-known, however, that as bandwidth-delay product continues to grow, TCP Reno will eventually become a performance bottleneck itself. The following four difficulties contribute to the poor performance of TCP Reno in networks with large bandwidth-delay products:

1)   At the packet level, linear increase by one packet per Round-Trip Time (RTT) is too slow, and multiplicative decrease per loss event is too drastic.

2)   At the flow level, maintaining large average congestion windows *requires* an extremely *small* equilibrium loss probability.

3)   At the packet level, oscillation is unavoidable because

      TCP uses a binary congestion signal (packet loss).

3

4) At the flow level, the dynamics is unstable, leading to severe oscillations that can only be reduced by the accurate estimation of packet loss probability and a stable design of the How dynamics.

We explain these difficulties in detail in Section II. In our project, we motivate delay-based approach. Delay-based congestion control has been proposed. Its advantage over loss-based approach is small at low speed, but decisive at high speed, as we will argue below. As pointed out in. delay can be a poor or untimely predictor of packet loss and therefore using a delay-based algorithm to augment the basic AIMD (Additive Increase Multiplicative Decrease) algorithm of TCP Reno is the wrong approach to address the above difficulties at large windows. Instead, a new approach that fully exploits delay as a congestion measure, augmented with loss information, is needed. FAST TCP uses this approach. Using queuing delay as the congestion measure has two advantages.

First, queuing delay can be more accurately estimated than loss probability both because packet losses in networks with large bandwidth-delay product are rare events (probability on the order 10^{-8} or smaller), and because loss samples provide coarser in formation than) queuing delay samples. Indeed, measurements of delay are noisy, just as those of loss probability. Each measurement of packet loss (whether a packet is lost) provides one bit of information for the filtering of noise. Whereas each measurement of queuing delay provides multi-bit information, this makes it easier for the equation-based implementation to stabilize a network into a steady state with a target fairness and high utilization. Second, the dynamics of queuing delay seems to have the right scaling with respect to network capacity. This helps maintain stability as a network scales up in capacity. In Section III we explain how we exploit these advantages to address the four difficulties of TCP Reno.

In Section IV, we lay out architecture to implement our design; Even though the discussion is in the context of FAST TCP the architecture can also serve as a general framework to guide the design of other congestion control mechanisms. Not necessarily limited to TCP, for high-speed networks. The main components in the architecture can be designed separately and upgraded asynchronously. Unlike the conventional design, FAST TCP can use the same window and burstiness control algorithms regardless of whether a source is in the normal state or the loss recovery state. This leads to a clean separation of components in both functionality and code structure. We then present an overview of some of the algorithms implemented in our current prototype.

In Section V, we present a mathematical model of the window control algorithm. We prove that FAST TCP has the same equilibrium properties as TCP Vegas. In particular, it does not penalize flows with large propagation delays, and it achieves weighted proportional fairness. For the special case of single bottleneck link with heterogeneous flows, we prove that the window control algorithm of FAST is globally stable, in the absence of feedback delay. Moreover, starting from any initial state, a network converges exponentially to a unique equilibrium.

In Section VI, we present the results to illustrate throughput, fairness, stability, and responsiveness of FAST TCP, in the presence of delay and in heterogeneous and dynamic environments where flows of different delays join and depart asynchronously. We compare the performance of FAST TCP with Reno, HSTCP (High-speed TCP, and STCP (Scalable TCP), using their default parameters. In these experiments, FAST TCP achieved the best performance under each criterion, while HSTCP and STCP improved throughput and responsiveness over Reno at the cost of fairness and stability. We conclude in Section VII.

5

II. PROBLEMS AT LARGE WINDOWS:

A congestion control algorithm can be designed at two levels. The flow-level (macroscopic) design aims to achieve high utilization, low queuing delay and loss, fairness, and stability. The packet - level design implements these (low- level goals within the constraints imposed by end-to-end control. Historically for TCP Reno, packet-level implementation was introduced first. The resulting flow -level properties, such as fairness, stability, and the relationship between equilibrium window and loss probability, were then understood as an afterthought. In contrast, the packet-level designs of HSTCP, STCP, and FAST TCP arc explicitly guided by flow-level goals.

We elaborate in this section on the four difficulties of TCP Reno listed in Section I. It is important to distinguish between packet-level and flow- level difficulties because they must be addressed by different means.

A .Packet and flow level modeling

The congestion avoidance algorithm of TCP Reno and its variants have the form of AIMD. The pseudo code for window adjustment is:

$$\textbf{Ack: w} \leftarrow \textbf{w+ (1/w)}$$

$$\textbf{Loss: w} \leftarrow \textbf{w-(1/w)}$$

This is a packet-level model, but it induces certain flow-level properties such as throughput, fairness, and stability. These properties can be understood with a flow-level model of the AIMD algorithm. The window of size increases by 1 packet per RTT and decreases per unit time by

$$x_i(t)p_i(t).(1/2). (4w_i(t)/3) \textbf{ packets}$$

where

$$x_i(t) = w_i(t)/T_i(t) \text{ packets/sec}$$

$T_i(t)$ is the round-trip time and $p_i(t)$ is the (delayed) end to end loss probability, in period t. Here $4w_i(t)/3$ is the peak window size that gives the "average" window of $w_i(t)$. Hence. a flow-level model of AIMD is:

$$w^*_i(t) = (1/T_i(t))-(2/3).x_i(t).p_i(t).w_i(t) \ldots\ldots\ldots\ldots(1)$$

Setting $w_i(t)= 0$ in yields the well known $1/\sqrt{p}$ formula for TCP Reno discovered, which relates loss probability to window size in equilibrium.

$$p^*_i = (3/(2w^*_i)^2) \ldots\ldots\ldots\ldots(2)$$

In summary (1) and (2) describe the flow-level dynamics and the equilibrium, respectively, for TCP Reno. It turns out that different variants of TCP all have the same dynamic structure at the flow level.

By defining

$$k_i(w_i, T_i) = (1/T_i) \ \& \ u_i(w_i, T_i) = 1.5/w_i^2.$$

and noting that

$$w_i(t) = k(t).(1 - (p_i(t)/u_i(t))\ldots\ldots\ldots\ldots(3)$$

where we have used the shorthand $k_i(t)= k_i(w_i(t), T_i(t))$ and $u_i(t)= u_i(w_i(t), T_i(t))$. Equation (3) can be used to describe all known TCP variants, and different variants differ in their choices of the gain function k_i and the marginal utility function u_i, and whether the congestion measure p_i is loss probability or queuing delay.

Next, we illustrate the equilibrium and dynamics problems of TCP Reno, at both the packet and flow levels, as bandwidth-delay product increases.

B. Equilibrium Problem:

The equilibrium problem at the flow level is expressed in (2): the end-to-end loss probability must be exceedingly small to sustain a large window size, making the equilibrium difficult to maintain in practice, as bandwidth-delay product increases.

Even though equilibrium is in flow-level notion, this problem manifests itself at the packet level, where a source increments its window too slowly and decrements it too drastically. When the peak window is 80,000-packct (corresponding to an "average" window of 60,000 packets), which is necessary to sustain 7.2Gbps using 1,500-byte packets with a RTT of l00ms, it takes 40,000 RTTs or almost 70 minutes, to recover from a single packet loss. This is illustrated in Figure 1a, where the size of window increment per RTT and decrement per loss. 1 and $0.5w_i$, respectively, are plotted as functions of w_i. The increment function for Reno (and for HSTCP) is almost indistinguishable from the x axis. Moreover, the gap between the increment and decrement functions grows rapidly as w_i increases. Since the average increment and decrement must be equal in equilibrium, the required loss probability can be exceedingly small at large w_i. This picture is thus simply a visualization of (2).To address the difficulties of Reno at large window sizes, HSTCP and STCP increase more aggressively and decrease more gently.

C. Dynamic Problems:

The causes of the oscillatory behavior of TCP Reno lie in its design at both the packet and flow levels. At the packet level, the choice of binary congestion signal necessarily leads to oscillation, and the parameter setting in Reno worsens the situation as bandwidth-delay product increases. At the flow level, the system dynamics given by (I) is unstable at large bandwidth-delay products. These must be addressed by different means, as we now elaborate.

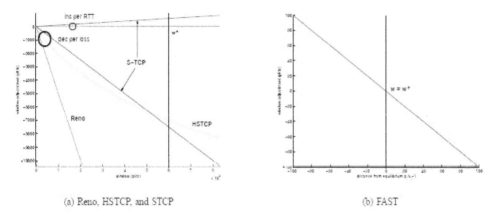

(a) Reno, HSTCP, and STCP

(b) FAST

Fig 1.(a).Window increment per RTT and decrement per loss as a function of current window in Packet level Implementation.

(b)Window update as a function of distance from equilibrium in FAST TCP

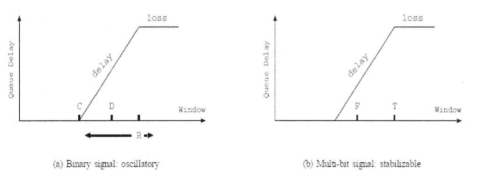

(a) Binary signal: oscillatory

(b) Multi-bit signal: stabilizable

Fig 2:(a).Operating points of TCP algorithms :R-Reno,C-CARD,D-DUAL,

9

Figure 2(a) illustrates the operating points chosen by various TCP congestion control algorithms, using the single-link, single-flow scenario. It shows queueing delay as a function of window size. Queueing delay starts to build up after point C where window equals bandwidth-propagation-delay product, until point R where the queue overflows. Since Reno oscillates around point R, the peak window size goes beyond point R. The minimum window in steady slate is half of the peak window. This is the basis for the rule of thumb that bottleneck buffer should be at least one bandwidth-delay product: the minimum window will then be above point *and* buffer will not empty in steady state operation, yielding full utilization.

In the loss-based approach, full utilization, even if achievable, comes at the cost of severe oscillations and potentially large queueing delay. The DUAL scheme proposes to oscillate around point D, the midpoint between C and R when the buffer is half-full. DUAL increases congestion window linearly by one packet per RTT, as long as queueing delay is less than half of the maximum value, and decreases multiplicatively by a factor of I /8. When queueing delay exceeds half of the maximum value. The scheme CARD (Congestion Avoidance using Round-trip Delay) proposes to oscillate around point C through A1MD with the same parameter **(1, 1/8)** as DUAL, based on the ratio of round-trip delay and delay gradient, to maximize power. In all these schemes, the congestion signal is used as a binary signal, and hence congestion window *must* oscillate.

Congestion window can be stabilized only if multi-bit feedback is used. This is the approach taken by the equation-based algorithm, where congestion window is adjusted based on the estimated loss probability in an attempt to stabilize around a target value given by (2). Its operating point is T in Figure

2(b), near the overflowing point. This approach eliminates the oscillation due to packet-level AIMD, but two difficulties remain at the flow level.

First, equation-based control requires the explicit estimation of end-to-end loss probability. This is difficult when the loss probability is small. *Second*, even if loss probability can be perfectly estimated, Reno's flow dynamics, described by equation (1) leads to a feedback system that becomes unstable as feedback delay increases, and more strikingly, as network capacity increases. The instability at the flow level can lead to severe oscillations that can be reduced *only* by stabilizing the flow-level dynamics. We will return to both points in Section III.

III. Delay-Based Approach:

In this section we, motivate delay-based approach to address the four difficulties at large window sizes.

A. Motivation

Although improved loss-based protocols such as HSTCP and STCP have been proposed as replacements to TCP Reno, we showed in [5] that they don't address all four problems (Section I) of TCP Reno. To illustrate this, we plot the increment and decrement functions of HSTCP and STCP in Figure l(a) alongside TCP Reno. Both protocols upper bound **TCP Reno**: each increases more aggressively and decreases less drastically, so that the gap between the increment and decrement functions is narrowed. This means, in equilibrium, both HSTCP and STCP can tolerate larger loss probabilities than TCP Reno, thus achieving larger equilibrium windows. However, neither solves the dynamics problems at both the packet and the flow levels.

In [5], we show that the congestion windows in Reno, HSTCP and STCP all evolve according to:

11

$$w_i(t) = K_i(t).(1-(p_i(t)/u_i(t)))........(4)$$

where $k(t) := ki(wi(t),Ti(t))$ and $wi(t) := u_i(w_i(t),Ti(t))$. Moreover, the dynamics of FAST TCP also takes the same form. They differ only in the choice of the gain function $k_i(w_i,T_i)$,the marginal utility function $u_i(w_i,T_i)$, and the end-to-end congestion measure p_i. Hence at the flow level, there are only three design decisions:

• $k_i(w_i ,T_i)$: the choice of the gain function k_i determines the dynamic properties such as stability and responsiveness, but does not affect the equilibrium properties.

•$u_i(w_i , T_i)$: the choice of the marginal utility function u_i mainly determines equilibrium properties such as the , equilibrium rate allocation and its fairness.

•p_i : In the absence of explicit feedback, the choice of congestion measure p_i is limited to loss probability or queueing delay. The dynamics of p_i (t) is determined at links.

The design choices in Reno, HSTCP, STCP and FAST are shown in Table 1:

Table I:

	$k_i(w_i,T_i)$	$u_i(w_i , T_i)$	p_i
Reno	$1/T_i$	$1.5/w_i^2$	Loss probability
HSTCP	$0.16b(w_i)w_i^{0.80}$ $(2-b(w_i))T_i$	$0.08/w_i^{1.20}$	Loss probability
STCP	aw_i/T_i	ρ/w_i	Loss probability
FAST	γa_i	a_i / x_i	Queing delay

Table II:

The above choices produce equilibrium characterizations as shown below:

Reno	$x_i = (1/T_i).(a_i/p_i^{0.50})$
HSTCP	$x_i = (1/T_i).(a_i/p_i^{0.84})$
STCP	$x_i = (1/T_i).(a_i/p_i)$
FAST	$x_i = (a_i/p_i)$

This common model (4) can be interpreted as follows: the goal at the flow level is to equalize marginal utility $u_i(t)$ with the end-to-end measure of congestion $p_i(t)$. This interpretation immediately suggests an equation-based packet-level implementation where *both* the direction and size of the window adjustment $w_i(t)$ are based on the difference between the ratio $p_i(t)/u_i(t)$ and the target of 1. Unlike the approach taken by Reno, HSTCP, and STCP, this approach eliminates packet-level oscillations due to the binary nature of congestion signal. It however requires the *explicit* estimation of the end-to-end congestion measure $p_i(t)$.

Without explicit feedback, $p_i(t)$ can only be loss probability, as used in TFRC [34]. or queueing delay, as used in TCP Vegas [8] and FAST TCP. Queueing delay can be more accurately estimated than loss probability both because packet losses in networks with large bandwidth-delay products are rare events (probability on the order 10^{-8} or smaller, and because loss samples provide coarser information than queueing delay samples. Indeed, each measurement of packet loss (whether a packet is lost) provides one bit of information for the filtering of noise, whereas each measurement of queueing delay provides multi-bit information. This allows an equation-based implementation to stabilize a network into a steady state with a target fairness and high utilization.

At the flow level, the dynamics of the feedback system must be stable in the presence of delay, as the network capacity increases. Here, again, queueing delay has an advantage over loss probability as a congestion measure: the dynamics of queueing delay seems to have the right scaling with respecl to network capacity. This helps maintain stability as network capacity grows.

B. Implementation Strategy:

The delay-based approach, with proper flow and packet level designs, can address the four difficulties of Reno at large windows. First, by explicitly estimating how far the current state $p_i(t)/u_i(t)$ is from the equilibrium value of 1, our scheme can drive the system rapidly, yet in a fair and stable manner, toward the equilibrium. The window adjustment is small when the current state is close to equilibrium and large otherwise, independent of whether the equilibrium is as illustrated in fig B. This is in stark contrast to the approach taken by Reno,HSTCP and STCP, where window adjustment depends on just the current window size and is independent of where the current state is with respect to the target (compare Figures l(a) and (b)). Like the equation-based scheme in [34], this approach avoids the problem of slow increase and drastic decrease in Reno. as the nefuork scales up.

Second, by choosing a multi-bit congestion measure, this approach eliminates the packet-level oscillation due to binary feedback, avoiding Reno's third problem.

Third, using queueing delay as the congestion measure $p_i(t)$ allows the network to stabilize in the region below the overflowing point, around point **F** in Figure 2(b), when the buffer size is sufficiently large. Stabilization at this operating point eliminates large queueing delay and unnecessary packet loss. More importantly, it makes room for buffering "mice" traffic. To avoid the second problem in Reno, where the required equilibrium congestion measure (loss probability for Reno, and queueing delay here) is too small to practically estimate, the algorithm must adapt its parameter a_i with capacity to maintain small but sufficient queueing delay.

Finally, to avoid the fourth problem of Reno, the window control algorithm must be stable, in addition to being fair and efficient at the flow level. The use of queue ing delay as a congestion measure facilitates the design as queueing delay naturally scales with capacity [22], [23], [24].

The design of TCP congestion control algorithm can thus be conceptually divided into two levels:

1) At the flow level, the goal is to design a class of function pairs, $u_i(w_i, T_i)$ and $k_i(w_i, T_i)$ so that the feedback system described by (4), together with link dynamics in $p_i(t)$ and the interconnection, has an equilibrium that is fair and efficient, and that the equilibrium is stable, in the presence of feedback delay.

2) At the packet level, the design must deal with issues that are ignored by the flow-level model or modeling assumptions that are violated in practice, in order to achieve these flow-level goals. Those issues include burstiness control, loss recovery, and parameter estimation.

The implementation then proceeds in three steps:

1) Determine various system components;

2) Translate the flow-level design into packet- level algorithms.

3) Implement the packet- level algorithms in a specific operating system.

15

The actual process iterates intimately between flow and packet level designs, between theory, implementation, and experiments, and among the three implementation steps.

The emerging theory of large-scale networks under end-to-end control forms the foundation of the flow-level design. The theory plays an important role by providing a framework to understand issues, clarify ideas, and suggest directions, leading to a robust and high performance implementation.

We lay out the architecture of FAST TCP next.

IV. Architecture and Algorithms:

We separate the congestion control mechanism of TCP into four components in Figure 3. These four components are functionally independent so that they can be designed separately and upgraded asynchronously. In this section, we focus on the two parts that we have implemented in the current prototype.

Fig 3. Architecture of FAST TCP:

Data Control	Window Control	Burstiness Control
	Estimation	

TCP Protocol Processing

16

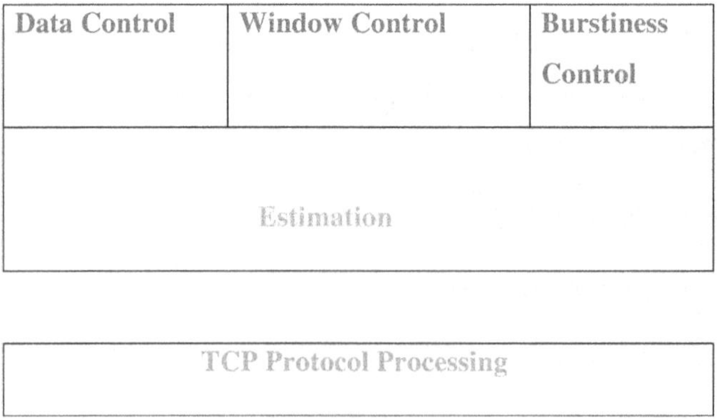

The data control component determines which packets to transmit, window control determines how many packets to transmit, and burstiness control determines when to transmit these packets. These decisions are made based on information provided by the estimation component. Window control regulates packet transmission at the RTT timescale, while burstiness control works at a smaller timescale. In the following subsections, we provide an overview of window control and algorithms implemented in our current prototype.

A. Estimation:

This component provides estimations of various input parameters to the other three decision-making components. It computes two pieces of feedback information for each data packet sent. When a positive acknowledgment is received, it calculates the RTT for the corresponding data packet and updates the average queueing delay and the minimum RTT. When a negative acknowledgment (signaled by three duplicate acknowledgments or timeout) is received, it generates a loss indication for this data packet to the other components. The estimation component generates both a multi-bit queueing delay sample and a one-bit loss-or-no loss sample for each data packet.

The queueing delay is smoothed by taking a moving average with the weight

$\eta(t) := \min\{3w_i(t), 1/4\}$ that depends on me window $w_i(t)$ at time t as follows. The k-th RTT sample $T_i(k)$ updates the average RTT $\mathbf{F}_i(\mathbf{k})$ according to:

$$\mathbf{F}_i(\mathbf{k+1}) = (1 - \eta(t_k)) \, \mathbf{F}_i(\mathbf{k}) + \eta(t_k) \, T_i(\mathbf{k})$$

Where t_k is the time at which the k-th RTT sample is received. Taking $d_i(k)$ to be the minimum RTT observed so far , the average queueing delay is estimated as

$$q_i(k) = \mathsf{T}_i(k) - d_i(k)$$

The weight $\eta(t)$ is usually much smaller than the weight (1/8) used in TCP Reno. The average RTT $\mathsf{T}_i(k)$ attempts to track the average over one congestion window. During each RTT an entire window worth of RTT samples are received if every packet is acknowledged. Otherwise, if delayed Ack is used, the number of queueing delay samples is reduced so $\eta(t)$ should be adjusted accordingly.

B. Window Control:

The window control component determines congestion window based on congestion information — queueing delay and packet loss, provided by the estimation component. A key decision in our design that departs from traditional TCP design is that the same algorithm is used for congestion window computation independent of the state of the sender. For example, in TCP Reno (without rate halving), congestion window is increased by one packet every RTT when there is no loss, and increased by one for each duplicate ack during loss recovery. In FAST TCP, we would like to use the same algorithm for window computation regardless of the sender

Our congestion control mechanism reacts to both queueing delay and packet loss. Under normal network conditions, FAST periodically updates the congestion window based on the average RTT and average queueing delay provided by the estimation component, according to (**5**):

18

w ← *min {2w, (1-γ)w* + *γ((base RTT /RTT)w* + *a(w, qdelay)}*............(5).

where $\gamma \in (0,1)$,base RTT is the minimum RTT observed so far, and qdelay is the end-to-end (average) queueing delay. In our current implementation, congestion window changes over two RTTs: it is updated in one RTT and frozen in the next. The update is spread out over the first RTT in a way such that congestion window is no more than doubled in each RTT.

In our current prototype, we choose the function **a(w, qdelay)** to be a constant at all times. This produces linear convergence when the qdelay is zero. Alternatively, we can use a constant **a** only when qdelay is non zero and an **a** proportional to window. **a(w, qdelay)** = **aw**. In this case when qdelay is zero FAST performs multiplicative increase and grows exponentially at rate **a** to a neighborhood of qdelay >0. Then **a(w, qdelay)** switches, to a constant **a** and, as we will see in Theorem 2 below, window converges exponentially to the equilibrium at a different rate that depends on qdelay. The constant **a** is the number of packets each flow attempts to maintain in the network buffer(s) at equilibrium, similar to TCP Vegas.

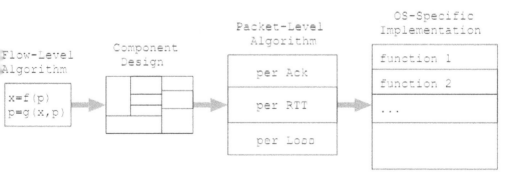

Fig 4. From Flow Level Design to Implementation.

Although we would like to use the same congestion control function during loss recovery, we have currently disabled this feature because of ambiguities associated with retransmitted packets. Currently when a packet loss is detected, FAST halves its window and enters loss recovery. The goal is to back off packet transmission quickly when severe congestion occurs, in order to bring the system back to a regime where reliable RTT measurements are again available for window adjustment (5) to work effectively. A source does not react to delay until it exits loss recovery.

C. Packet- Level Implementation:

It is important to maintain an abstraction as the code evolves. This abstraction should describe the high-level operations each component performs based on external inputs, and can serve as a road map for future TCP implementations as well as improvements to the existing implementation. Whenever a non-trivial change is required, one should first update this abstraction to ensure that the overall packet-level code would he built on a sound underlying foundation.

Since TCP is an event-based protocol, our control actions should be triggered by the occurrence of various events. Hence, we need to translate our flow-level algorithms into event-based packet-level algorithms. There are four types of events that FAST TCP reacts to: on the reception of an acknowledgment, after the transmission of a packet, at the end of a RTT, and for each packet loss.

For each acknowledgment received, the estimation component computes the average queueing delay, and the burstiness control component determines whether packets can be injected into the network. For each packet transmitted, the estimation component records a time-stamp, and the burstiness control component updates corresponding data structures for book-keeping. At a constant time interval, which we check on the arrival of each acknowledgment, window control calculates a new window size. At the end of each RTT, burstiness reduction calculates the target throughput using the window and RTT measurements in the last RTT. Window pacing will then schedule to break up a large increment in congestion window into smaller increments over time. During loss recovery, congestion window should be continually updated based on congestion signals from the network. Upon the detection of a packet loss event, a sender determines whether to retransmit each un-acknowledged packet right away or hold off until a more appropriate time.

Figure 4 presents an approach to turn the high level design of a congestion control algorithm into an implementation. First, an algorithm is designed at the flow-level and analyzed to ensure that it meets the high-level objectives such as fairness and stability. Based on that, one can determine the components necessary to implement congestion control.

The flow-level algorithm can then be translated into a packet-level algorithm that consists of a set of event-based tasks. The event-based tasks should be independent of any specific TCP or operating system implementation, but yet detailed enough so the understanding of these tasks enables one to implement FAST in any operating system or protocol stack.

V. Equilibrium and Stability Of Window Control Algorithm:

In this section we present a model of the window control algorithm. We show that in equilibrium , the vectors of source windows and link queueing delays are the unique solutions of pair of optimization problems(9)—(10). This completely characterizes the network equilibrium properties such as throughput,

21

fairness, and delay. We also analyze the stability of the window control algorithm. We prove that, for a single link with heterogeneous sources, the window control algorithm (5) is globally stable, assuming zero feedback delay, and converges exponentially to a unique equilibrium. Extensive experiments in Section VI illustrate this stability in the presence of feed back delay.

Given a network that consists of a set of resources with finite capacities c_l. e.g. transmission links, processing units, memory etc. We refer to them in general as "links" in our model. The network is shared by a set of unicast flows identified by their sources. Let d_i denote the round-trip propagation delay of source i. Let R be the routing matrix where $R_{li} = 1$ if source **i** uses link **f**, and 0 otherwise. Let $q_i(t) = \Sigma_l\ R_{li}\ pi(t)$denote the round trip queueing delay *or* in vector notatin $q_i(t)=R^Tp(t)$. Then the RTT of source *i* is $T_i(t)=d_i+q_i(t)$

Each source I adapts $w_i(t)$ periodically according to

$$w_i(t+1) = \gamma((d_iw_i(t)/(d_i+q_i(t))+a_i(w_i(t),q_i(t))+(1-\gamma)w_i(t)\ldots\ldots\ldots(6)$$

where $\gamma \in (0,1)$, at time *t*, and $a_i(wi,q_i)$is defined by:

$$a_i(w_{i,}q_i) = \{a_iw_i\ if\ q_i=0\ \ldots\ldots\ldots\ldots\ldots(7)$$

$$\{a_i,\ \ otherwise$$

A key departure in our model from those in the literature is that we assume that a source's send rate defined as $x_i(t)=w_i(t)/T_i(t\ )$,cannot exceed the through put it receives .This is justified because of self-clocking: one round-trip time after a congestion window is increased, packet transmission will be clocked at the same rate as the throughput the flow receives. A consequence of this assumption is that the link queueing delay vector, *p(t),* is determined implicitly by die instantaneous window size in a static manner: given $wi(t)=$

w_i for all i, the link queueing delays $p_l(t) = p_l(t) > 0$ for all l are given by:

$$\Sigma\, R_{li}.(w_i/(d_i+q_i))\{=c_i \quad \text{if } p_l{>}0 \ \dots\dots\dots\dots\dots(8)$$

$$\{<c_i \quad \text{if } p_l{=}0$$

Where again $q_i = \Sigma\, R_{li}\, p_l$

The equilibrium values of windows w^* and delays p^* of the network defined by (6) --- (8) can be characterized as follows. Consider the utility maximization problem

$$Max\, \Sigma\, a_i log x_i\ s.t.\ R_r < c, \dots\dots\dots\dots\dots(9)$$

$$x{>}0 \quad i$$

and the following dual problem

$$min\ \Sigma\, c_l p_l - \Sigma a_i log\, \Sigma R_{li} p_L \dots\dots\dots\dots\dots(10)$$

$$p{>}0 \quad i \qquad i \qquad i$$

Theorem 1:

Suppose R has full row rank. The unique equilibrium point (w^*,p^*) of the network is defined by (6)—(8) exists and is such that $x^* = (x_i^* = w_i/(d_i+q_i^*), \forall i)$ is the unique maximiser of (9) and p^* is the unique minimiser of (10).This implies in particular that the equilibrium rate x^* is ai – weighted proportionally fair.

Theorem 1 implies that FAST TCP has the same equilibrium properties as TCP Vegas. It s throughput is given by

$$x_i = a_i /\, q_i. \dots\dots\dots\dots(11)$$

In particular it does not penalize sources with large propogation delays d_i The relation(11) also implies that in equilibrium source I maintains a_i packets in the buffers along its path. Hence the total amount off buffering in the network must be atleast $\Sigma_i a_i$ packets inorder to reach the equilibrium.

We now turn to the stability of the algorithm. Global stability in a general network in the presence of feedback delay is an open problem . State-of-the-art results either prove global stability while ignoring feedback delay, or local stability in the presence of feedback delay. Our stability result is restricted to a single link in the absence of delay.

Theorem 2:

Suppose there is a single link with capacity c. Then the network defined by (6)-(8) is globally stable, and converges geometrically to the unique equilibrium (w^*, p^*).

The basic idea of the proof is to show that the iteration from $w(t)$ to $w(t + 1)$ defined by (6)--(8) is a contraction mapping. Hence $w(t)$ converges geometrically to the unique equilibrium.

Some properties follow from the proof of Theorem 2.

Corollary 3:

1) Starting from an initial point $(w(0), p(0))$the link is fully utilized, i.e.. equality holds in (8), after a finite time.

2) The queue length is lower and upper bounded after a finite amount of time.

VI. PERFORMANCE:

We have conducted some preliminary experiments on our dummynet testbed kernel comparing performance of various new TCP algorithms as well as the Linux TCP implementation. It is important to evaluate them not only in static environments, but also dynamic environments where flows come and go; and not only in terms of end-to-cnd throughput, but also queue behavior in the network. In this study, we compare performance among TCP connections of the same protocol sharing a single bottleneck link. In summary,

1)   FAST TCP achieved the best overall performance in each of the four evaluation criteria: throughput, fairness, responsiveness, and stability.

2) Both HSTCP and STCP improved throughput and responsiveness of Linux TCP, although both showed fairness problems and oscillations with higher frequencies and larger magnitudes.

A. Testbed And Kernel Implementation:

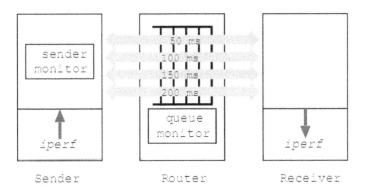

Fig 5: Testbed and Kernel Setup.

Figure 5 shows the setup of the testbed. The testbed router supports paths of various delays and a single bottleneck capacity with a fixed buffer size. It has monitoring capability at the sender and the router. The receiver runs different TCP traffic sinks with different port numbers for connections with different RTTs. We set up and run different experiments from the sender using an automatic script generator to start multiple *iperf* sessions to emulate multiple TCP connections.

Our testbed router ran dummynet under Free-BSD. We configured dummynet to create paths or pipes of different delays, 50, 100, 150, and 200ms, using different destination port numbers on the receiving machine. We then created another pipe to emulate a bottleneck capacity of 800 Mbps and a buffer size of 2,000 packets, shared by all the delay pipes. Due to our need to emulate a high-speed bottleneck capacity, we increased the scheduling granularity of dummynet events. We recompiled the FreeBSD kernel so the task scheduler ran every 1ms. We also increased the size of the IP layer interrupt queue *(ipintrq)* to 3000 to accommodate large bursts of packets.

We instrumented both the sender and the dummynet router to capture relevant information for protocol evaluation. For each connection on the sending machine, the kernel monitor captured the congestion window, the observed base RTT, and the observed queueing delay. On the dummynet router, the kernel monitor captured the throughput at the dummynet bottleneck, the number of lost packets, and the average queue size every two seconds. We retrieved the measurement data after the completion of each experiment in order to avoid disk I/O that may have interfered with the experiment itself.

We tested four TCP implementations: FAST, HSTCP, STCP, and Reno (Linux implementation). The FAST TCP is based on Linux 2.4.20 kernel, while the rest of the TCP protocols are based on Linux 2.4.19 kernel. We ran tests and did not observe any appreciable difference between the two plain Linux kernels, and the TCP source codes of the two kernels are nearly identical. Linux TCP implementation includes all of the latest RFCs such as New Reno, SACK, D-SACK, and TCP high performance extensions. There are two versions of HSTCP.

In all of our experiments, the bottleneck capacity is 800 Mbps --roughly 66 packets/ms, and the maximum buffer size is 2000 packets.

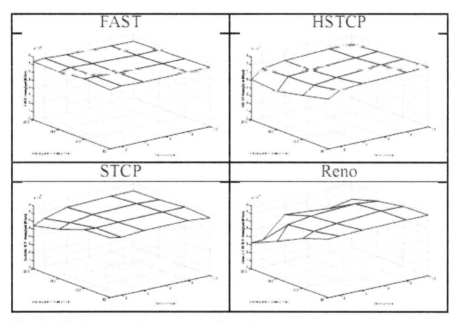

Fig. 6. Static scenario: aggregate throughput.

We now present our experimental results. We first look at three cases in detail, comparing not only the throughput behavior seen at the source, but also the queue behavior inside the network, by examining trajectories of throughputs, windows, instantaneous queue, cumulative losses, and link utilization. We then summarize the overall performance in a diverse set of experiments in terms of quantitative metrics, defined below, on throughput, fairness, stability, and responsiveness.

B. Case study: static scenario:

We present experimental results on aggregate throughput in a simple static environment where, in each experiment, all TCP flows had the same propagation delay and started and terminated al the same times. This set of tests included 20 experiments for different pairing of propagation delays. 50. 100, 150, and 200ms, and the number of identical sources, 1, 2, 4, 8, and 10, we ran this test suite under each of the four TCP protocols. We then constructed a 3-d plot, in Figure 6. For each protocol with the x-axis and y-axis being tie number of sources and propagation delay, respectively. The z-axis is the aggregate throughput.

All four protocols performed well when the number of flows was large or the propagation delay was small, i.e., when the window size was small. The performance of FAST TCP remained consistent when these parameters changed. TCP Reno, HSTCP and STCP had varying degrees of performance degradation as the window size increased, with TCP Reno showing the most significant degradation.

This set of experiments, involving a static environment and identical flows, does not test the fairness stability and responsiveness of the protocols. We take a close look at these properties next in a dynamic scenario where network equilibrium changes as flows come and go.

C. Case study: dynamic scenario I:

In the first dynamic test, the number of flows was small so that throughput per flow, and hence the window size, was large. There were three TCP flows, with propagation delays of 100, 150, and 200ms that started and terminated at different times, as illustrated in Figures 7(a).

For each dynamic experiment, we generated two set of figures, from the sender monitor and the queue monitor. From the sender monitor, we obtained the trajectories of individual connection throughput (in Kbps) and window size (in packets) over time. They are shown in Figure 8.

As new flows joined or old flows left, FAST TCP converged to the new equilibrium rate allocation rapidly and stably. Reno's throughput was also relatively smooth because of the slow (linear) increase before packet losses. The link utilization was low at the end of the experiment when it took 30 minutes for a flow to reclaim the spare capacity due to the departure of another flow. HSTCP and STCP, in an attempt to respond more quickly, went into severe oscillation.

From the queue monitor, we obtained three trajectories: the average queue size (packets), the number of cumulative packet losses(packets), and the utilization of the bottleneck link (in packets/ms), shown in Figure 9 from top to bottom. The queue under FAST TCP was quite small throughout the experiment due to the small number of flows. HSTCP and STCP exhibited strong oscillations that filled the buffer. The link utilizations of FAST TCP and Reno were quite steady, whereas those of HSTCP and STCP showed fluctuations.

From the throughput trajectories of each protocol, we calculate Jain's fairness index (sec Section VI-E for definition) for the rate allocations for each time interval that contains more than one flow (sec Figure 7(a)). The fairness indices are shown in Table III. FAST TCP obtained the best fairness very close to 1, followed by HSTCP, Reno, and then STCP confirms that FAST TCP does not penalize against flows with large propagation delays.

Time	#Sources	FAST	HSTCP	STCP	Reno
1800 – 3600	2	.967	.927	.573	.684
3600 – 5400	3	.970	.831	.793	.900
5400 – 7200	2	.967	.873	.877	.718

TABLE III
DYNAMIC SCENARIO I: INTRA-PROTOCOL FAIRNESS.

Even though HSTCP, STCP, and Reno all try to equalize congestion windows among competing connections instead of equalizing rates, this was not achieved as shown in Figure 8. The unfairness is especially severe in the case of STCP, likely due to MIMD. For FAST TCP, each source tries to maintain the same number of packets in the queue in equilibrium, and thus, in theory, each competing source should get an equal share of the bottleneck bandwidth. Even though FAST TCP achieved the best fairness index, we did not observe the expected equal sharing of bandwidth (see Figure 8). We found that, connections with longer RTTs consistently observed higher queueing delays than those with shorter RTTs. For example, the connection on the path of 100 ms saw an average queueing delay of 6 ms, while the connection on the path of 200 ms saw an average queueing delay of 9 ms. This caused the connection with longer RTTs to maintain fewer packets in the queue in equilibrium, thus getting a smaller share of the bandwidth. We have yet to uncover the source of this problem, but the early conjecture is that when congestion window size is large, it is much harder to break up bursts of packets. With bursty traffic arriving at a queue, each packet would see a delay that includes the transmission times of all preceding packets in the burst. However, if packets were *spaced out smoothly*, then each packet would have seen a smaller queueing delay at the queue.

D. Case Study: Dynamic scenario II:

This experiment was similar to dynamic scenario I, except that there were a larger number of flows, with different propagation delays, which joined and departed according to the schedule in Figure 7(b). The qualitative behavior in throughput, fairness, stability, and responsiveness for each of the protocols is similar to those in scenario I, and in fact is amplified as the number of flows increases.

Specifically, as the number of competing sources increases in a network, stability becomes worse for the loss-based protocols. As shown in Figures 10 and 11, oscillations in both congestion windows and queue size are more severe for all loss-base protocols. Packet loss is also more severe. The performance of FAST TCP did not degrade in any significant way. Connections sharing the link achieved very similar rates. There was a reasonably stable queue at all times, with little packet loss and high link utilization. Intra-protocol fairness is shown in Table IV, with little change for FAST TCP.

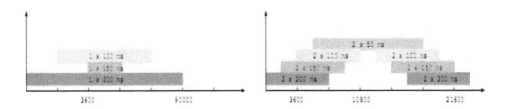

Fig 7: dynamic scenario implementation, each colored block represents one or more connections of certain propagation delay with the left and right edges representing the starting and ending times respectively

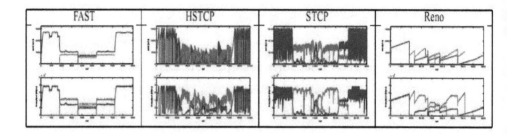

Fig 8: Throughput and congestion window trajectories-theoretical depiction

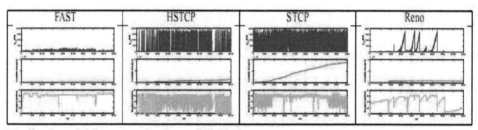

Fig. 9. Dynamic scenario I: dummynet queue sizes, losses, and link utilization.

Fig 9: Dummynet queue sizes, losses and link utilization

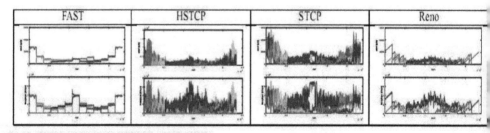

Fig. 10. Dynamic scenario II: throughput and cwnd trajectories.

Fig 10: Dynamic scenario II: Throughput and Congestion window trajectories.

32

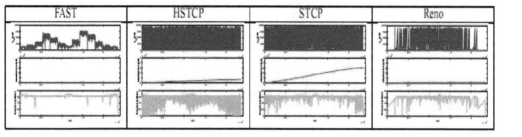

FAST	HSTCP	STCP	Reno

Fig 11: Dummynet queue sizes losses and link utilization for Dynamic scenario II

E. Overall evaluation

We use the output of *iperf* for our quantitative evaluation. Each *iperf* session in our experiments produced five-second averages of its throughput. This is the data rate (i.e., goodput) applications such as iperf receives, and is slightly less than the bottleneck bandwidth due to IP and Ethernet packet headers.

Let $x_i(k)$ be the average throughput of flow i in the five-second period k. Most tests involved dynamic scenarios where flows joined and departed. For the definitions below, suppose the composition of flows changes in period *k=1…..m, and changes again* over period $k = m+1$ so that [1, m] is the maximum-length interval over which the- same equilibrium holds. Suppose there are n active flows in this interval, indexed by i=1……,n.

Let

$$\overline{x}_i \; := \; \frac{1}{m} \sum_{k=1}^{m} x_i(k)$$

be the average throughput of flows *i* over this interval. We now define our performance metrics for this interval *[1,m]* using these throughput measurements.

1)Throughput: The average aggregate throughput for the interval [l,m] is defined as:

$$E := \sum_{i=1}^{n} \overline{x}_i$$

2)Intra-protocol fairness: Jain's fairness index for the interval *[1,m]* is defined as

$$F = \frac{\left(\sum_{i=1}^{n} \overline{x}_i\right)^2}{n \sum_{i=1}^{n} \overline{x}_i^2}$$

F ∈ (0,1) and *F = 1* is ideal (equal sharing).

3) Stability: The stability index of flow *i* is the sample standard deviation normalized by the average throughput:

$$S_i := \frac{1}{\overline{x}_i} \sqrt{\frac{1}{m-1} \sum_{k=1}^{m} (x_i(k) - \overline{x}_i)^2}$$

The smaller the stability index, the less oscillation a source experiences. The stability index for interval [0, m] is the average over the *n* active sources:

$$S := \frac{1}{n} \sum_{i=1}^{n} S_i$$

4) *Responsiveness:* The responsiveness index measures the speed of convergence when the network equilibrium changes at k=1, i.e. when flows join or depart. Let $x_i(k)$ be the running average by period $k < m$:

$$\overline{x}_i(k) \quad := \quad \frac{1}{k}\sum_{t=1}^{k} x_i(k)$$

For each TCP protocol, we obtain one set of computed values for each evaluation criterion for all of our experiments. We plot the CDF (cumilative distribution function) of each set of values. These are shown in Figures 12 - 15.

From Figures 12- 15, FAST has the best performance among all protocols under each evaluation criterion. More importantly, the variation in each of the distributions is smaller under FAST than under the other protocols, suggesting that FAST had fairly consistent performance in our test scenarios. We also observe that both HSTCP and STCP achieved higher throughput and improved responsiveness compared with TCP Reno.

STCP had worse intra-protocol fairness compared with TCP Reno, while HSTCP achieved comparable intra-protocol fairness to Reno (see Figures 13 and 10). Both HSTCP and STCP showed increased oscillations compared with Reno (Figures 14. 8 and 9), and the oscillations became worse as the number of sources increased (Figures 10 and 11),

35

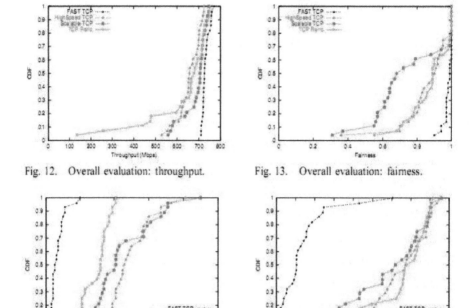

Fig. 12. Overall evaluation: throughput.

Fig. 13. Overall evaluation: fairness.

Fig. 14. Overall evaluation: stability.

Fig. 15. Overall evaluation: responsiveness index R_1.

From Figure 15, FAST TCP achieved a much better responsiveness index R (which is based on worst case individual throughput) than the other schemes. We caution however that it can be hard to quantify "responsiveness" for protocols that do not stabilize into an equilibrium point or a periodic limit cycle, and hence the unresponsiveness of Reno, HSTCP and STCP, as measured by index R should be interpreted with care.

F. Torrents –A real-time application presently using TCP download:

Torrent is a peer-to-peer file sharing protocol used for distributing large amounts of data. Bit Torrent is one of the most common protocols for transferring large files, and by some estimates it accounts for about 35% of all traffic on the entire Internet. The protocol works initially when a file provider makes his file (or group of files) available to the network. This is called a *seed* and allows others, named *peers*, to connect and download the file. Each peer who downloads a part of the data makes it available to other peers to download. After the file is successfully downloaded by a peer, many continue to make the data available, becoming additional seeds. This distributed nature of Bit Torrent leads to a viral spreading of a file throughout peers. As more seeds get added, the likelihood of a successful connection increases exponentially. Relative to standard Internet hosting, this reduces the original distributor's hardware and bandwidth resource costs. It also provides redundancy against system problems and reduces dependence on the original distributor. Programmer Bram Cohen designed the protocol in April 2001 and released a first implementation on July 2, 2001. It is now maintained by Cohen's company Bit Torrent, Inc. there are numerous Bit Torrent clients available for a variety of computing platforms. According to isoHunt, the total amount of shared content is currently more than 1.1 petabytes. A Bit Torrent client is any program that implements the Bit Torrent protocol. Each client is capable of preparing, requesting, and transmitting any type of computer file over a network, using the protocol. A peer is any computer running an instance of a client. To share a file or group of files, a peer first creates a small file called a "torrent" (e.g. MyFile.torrent). This file contains metadata about the files to be shared and about the tracker, the computer that coordinates the file distribution. Peers that want to download the file must first obtain a torrent file for it, and connect to the specified tracker, which tells them from which other peers to download the pieces of the file.

Though both ultimately transfer files over a network, a Bit Torrent download differs from a classic full-file HTTP request in several fundamental ways:

- Bit Torrent makes many small data requests over different TCP sockets, while web browsers typically make a single HTTP GET request over a single FTP socket.

- Bit Torrent downloads in a random or in a rarest-first approach that ensures high availability, while HTTP downloads in a sequential manner.

Taken together, these differences allow Bit Torrent to achieve much lower cost to the content provider, much higher redundancy, and much greater resistance to abuse or to "flash crowds" than a regular HTTP server. However, this protection comes at a cost: downloads can take time to rise to full speed because it may take time for enough peer connections to be established, and it takes time for a node to receive sufficient data to become an effective up loader. As such, a typical Bit Torrent download will gradually rise to very high speeds, and then slowly fall back down toward the end of the download. This contrasts with an HTTP server that, while more vulnerable to overload and abuse, rises to full speed very quickly and maintains this speed throughout.

G. Coding for FAST TCP in NS2:

Tcp-fast.h module:

```
#ifndef ns_tcp_fast_h
#define ns_tcp_fast_h


#include "tcp.h"
#include "flags.h"
#include "scoreboard.h"
#include "scoreboard-rq.h"
#include "random.h"
#include <stdio.h>
#include <stdlib.h>
#include <sys/types.h>


#undef FASTTCPAGENT_DEBUG
#define UPDATE_CWND_EVERY_OTHER_RTT


class FastTcpAgent : public virtual TcpAgent {
public:
      FastTcpAgent();
      ~FastTcpAgent();
```

virtual void recv(Packet *pkt, Handler*);

virtual void reset();

void   fast_est(Packet *pkt,double rtt);

void   fast_cc(double rtt, double old_pif);

double   fast_calc_cwnd(double cwnd, double old_pif);

void   fast_recv_newack_helper(Packet *pkt);

double   fast_est_update_avgrtt(double rtt);

double   fast_est_update_avg_cwnd(int inst_cwnd);

void   fast_pace(TracedDouble *cwndp, int incre_4_cwnd);

int   fast_expire(Packet* pkt);

virtual void timeout(int tno);

virtual void output(int seqno, int reason);

#ifdef FASTTCPAGENT_DEBUG

FILE * fasttcpagent_recordfps[10];

#endif

protected:

virtual void delay_bind_init_all();

virtual int   delay_bind_dispatch(const char *varName, const char *localName, TclObject *tracer);

virtual void traceVar(TracedVar* v);

virtual void dupack_action();

virtual void partial_ack_action();

virtual void send_much(int force, int reason, int maxburst);

TracedInt   alpha_;

TracedInt   beta_;

TracedDouble mi_threshold_;

TracedDouble   avgRTT_;

TracedDouble   avg_cwnd_last_RTT_;

TracedDouble   baseRTT_;

unsigned char   fast_opts;

int   alpha_tuning_;

TracedInt   high_accuracy_cwnd_;

double   fast_calc_cwnd_end;

int   slowstart_;   // # of pkts to send after slow-start,
deflt(2)

bool   on_first_rtt_;   //use this to identify if FAST should freeze or
update its cwnd in this rtt.

double cwnd_remainder; //this variable is used to fix the calculation error
using old_cwnd.

double   cwnd_update_time;

```
double              fast_update_cwnd_interval_;

double              cwnd_increments;

unsigned int        acks_per_rtt;

unsigned int        acks_last_rtt;

unsigned short      bc_ack,bc_spacing;

double rtt_;              // current rtt;

double newcwnd_;         // record un-inflated cwnd

double* sendtime_;              // each unacked pkt's sendtime is
recorded.

                         // (fixes problems with RTT calc'n)

double*    cwnd_array_;

int*   transmits_;       // # of retx for an unacked pkt

int    maxwnd_;               // maxwnd size for v_sendtime_[]


double firstrecv_;       //first receive timestamp.

double currentTime;           //the time when a recv event happens


u_char timeout_;  /* boolean: sent pkt from timeout? */

u_char fastrecov_; /* boolean: doing fast recovery? */

TracedInt pipe_;        /* estimate of pipe size (fast recovery) */

int partial_ack_;  /* Set to "true" to ensure sending */
```

```
                              /*  a packet on a partial ACK.    */
        int next_pkt_;                /* Next packet to transmit during Fast */
                              /*  Retransmit as a result of a partial ack. */
        int firstpartial_;    /* First of a series of partial acks. */
        int last_natural_ack_number_;  /*added by me, the last 'natural' ack
number */
                              /* which is only updated when last_ack_ <
tcph->seqno()*/
        ScoreBoard* scb_;
        static const int SBSIZE=64; /* Initial scoreboard size */


        double gamma_;


        double fasttime() {
                return(Scheduler::instance().clock() - firstsent_);
//              return (NOW);
        }
};
#endif // ns_tcp_fast_h
```

Fast-test1.tcl:

```
# Create a FAST Tcp connection
```

```
proc create_tcp_fast { idx src dst start stop } {

        global tcps

        global ftps

        global ns

        global sinks

        global tcps_start

        global tcps_stop

        puts "New TCP connection";

        # set source agent and application

        set tcps($idx) [new Agent/TCP/Fast]

        $tcps($idx) set class_ 2

        $tcps($idx) set window_ 100000

        $tcps($idx) set alpha_ 200

        $tcps($idx) set beta_ 200

        $ns attach-agent $src $tcps($idx)

        set ftps($idx) [new Application/FTP]

        $ftps($idx) attach-agent $tcps($idx)


        # set sink

        set sinks($idx) [new Agent/TCPSink/Sack1]

        $ns attach-agent $dst $sinks($idx)

        $ns connect $tcps($idx) $sinks($idx)

        set tcps_start($idx) [expr $start]
```

```
        set tcps_stop($idx) [expr $stop]

        $ns at $start "$ftps($idx) start"

        $ns at $stop  "$ftps($idx) stop"

}


# Record Procedure

proc record {} {

        global tcps ntcps

        global f0

        global queueMoniF queueMoniB

        global tcps_start tcps_stop


        #get an instance of the simulator

        set ns [Simulator instance]

        #set the time afterwhich the procedure should be called again

        set time 0.005


        set now [$ns now]

        set queueF [$queueMoniF set pkts_]

        set queueB [$queueMoniB set pkts_]

        set rate 0

        for {set i 0} {$i < $ntcps} {incr i} {

                if { ($now > $tcps_start($i)) && ($now < $tcps_stop($i)) } {
```

```
                  set window     [expr [$tcps($i) set cwnd_]];

                  set rtt [expr [$tcps($i) set avgRTT_]];

                  set bRTT     [expr [$tcps($i) set baseRTT_]];

                  set old_window    [expr            [$tcps($i)            set
avg_cwnd_last_RTT_]];

                  set total_pkt_num [expr [$tcps($i) set t_seqno_]]

                  if { $rtt > 0 } {

                  set rate      [expr $window/$rtt]

                  }

                  if { $bRTT < 9999999 } {

                         puts $f0 "$i $now $queueF $queueB $rate $window
$rtt $bRTT $old_window $total_pkt_num"

                         }

                  }

            }


      # Re-schedule the procedure

      $ns at [expr $now+$time] "record"

}
# Define 'finish' procedure (include post-simulation processes)

proc finish {} {

      global ns

      global f0

      global f_all
```

```
        $ns flush-trace

        close $f_all

        close $f0

        puts "Finished."

        exit 0

}


# ------ SIMULATION SPECIFIC SCRIPT -----------


# The preamble

set ns [new Simulator]

$ns use-scheduler Heap

# Predefine tracing

set Name "fast-test1-record.dat";

set Q1 "fast-test1-queue1.dat";

set Q2 "fast-test1-queue2.dat";

set f0 [open $Name w]

set f1 [open $Q1 w]

set f2 [open $Q2 w]

set f_all [open nstrace.dat w]

#$ns trace-all $f_all
```

```
# Define the topology

set Cap 100

set Del 50

set r1 [$ns node]

set r2 [$ns node]

set r3 [$ns node]


$ns duplex-link $r1 $r2  [expr $Cap]Mbps [expr $Del]ms DropTail


set botqF [[$ns link $r1 $r2] queue]

set botqB [[$ns link $r2 $r1] queue]

$ns queue-limit $r1 $r2 2000000;

$ns queue-limit $r2 $r1 2000000;


# Create Sources and Start & Stop times

set ntcps    3;

set SimDuration 100;

create_tcp_fast 0 $r1 $r2 0  100

create_tcp_fast 1 $r1 $r2 20 80

create_tcp_fast 2 $r1 $r2 40 60


#$ns create-connection TCP $r1 TCPSink $r2 1

set starttime 0.0
```

set finishtime $SimDuration

Monitor the queue between source switch and destination switch

set queueMoniF [$ns monitor-queue $r1 $r2 $f1 0.1]

set queueMoniB [$ns monitor-queue $r2 $r1 $f2 0.1]

Run the Simulation

$ns at 0.0 "record"

$ns at $SimDuration "finish"

puts "Simulating $SimDuration Secs."

$ns run

H. Results Obtained:

Test1:

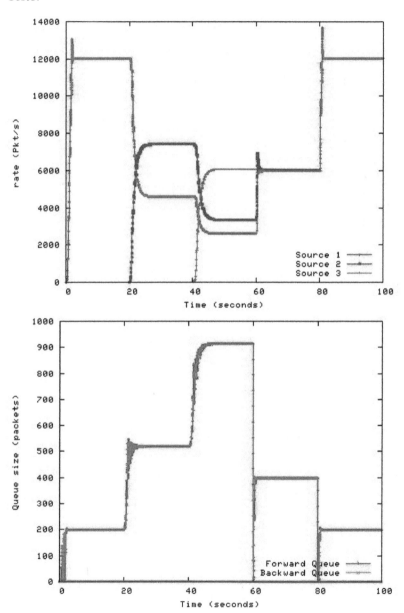

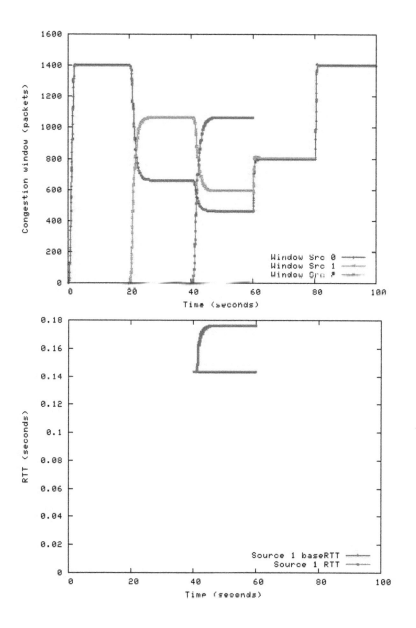

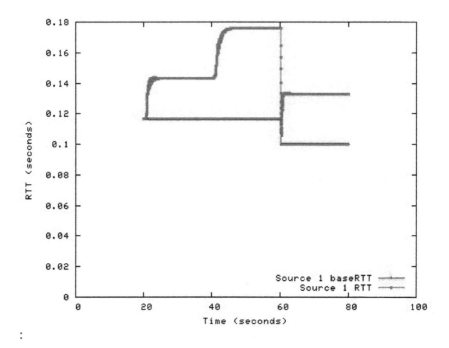

:

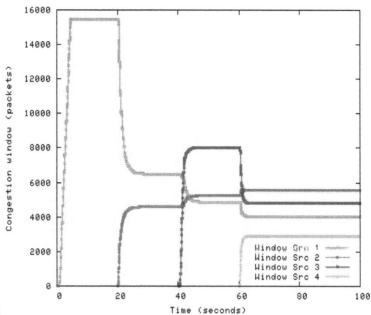

Test2:

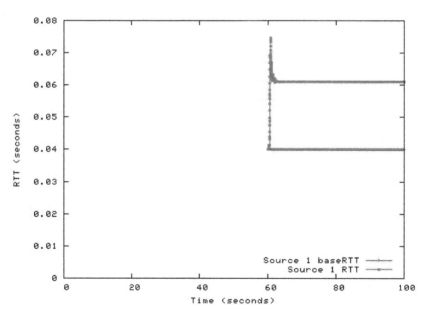

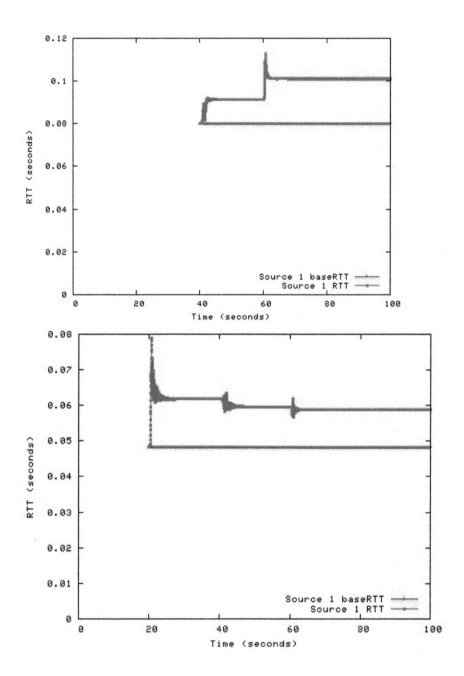

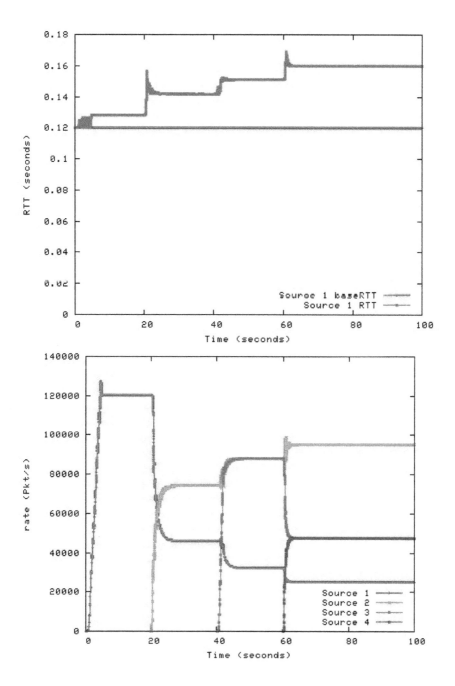

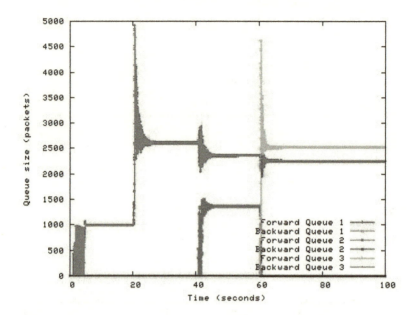

Test3:

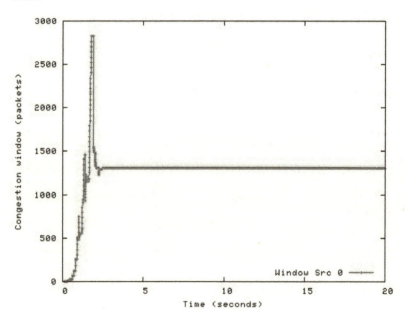

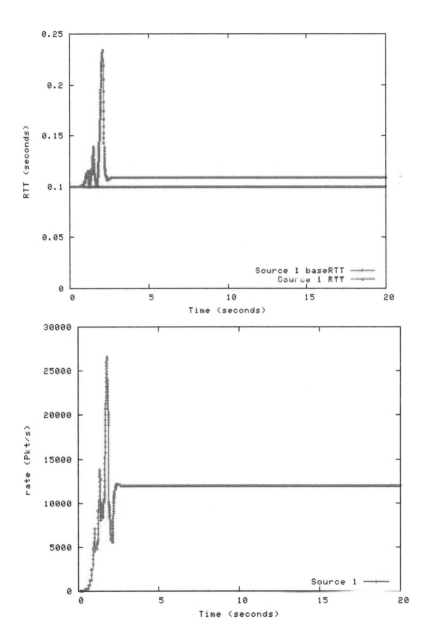

Test 4:

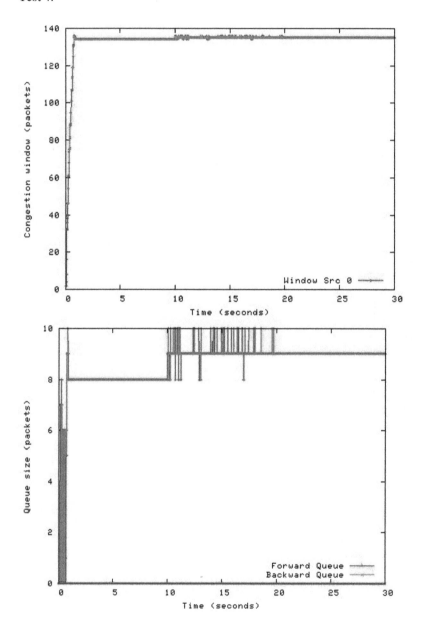

VII. Future Enhancement

The existing real time applications can be transformed to a large scale basis and can be implemented with ease. On enhancement of our algorithm, the new algorithm holds the key for many new frontiers to be explored in case of congestion control. The congestion control algorithm is currently running on Linux platform. The Windows platform is the widely used one. By proper Simulation applications, in Windows we can implement the same congestion control algorithm for Windows platform also. The Torrents application which we are currently using can achieve speeds similar to or better than "Rapid share (premium user)" application. Linux is also growing popularity nowadays. So our Project has outstanding scope in future.

VIII. Conclusion:

We have described and implemented an alternative congestion control algorithm, FAST TCP that addresses the four main problems of TCP Reno in networks with high capacities and large latencies. FAST TCP has a log utility function and achieves weighted proportional fairness. Its window adjustment is equation-based, under which the network moves rapidly toward equilibrium when the current state is far away and slows down when it approaches equilibrium. FAST TCP uses Queueing delay in addition to packet loss, as a congestion signal. Queueing delay provides a finer measure of congestion and scales naturally with network capacity. On proper implementation of our project, we can do many safe and fast downloads and data transfers over a high speed internet network.

REFERENCES

1. D. Chiu and R. Jain, "Analysis of the increase and decrease algorithms for congestion avoidance in computer networks," Computer Networks, vol. 17, pp. 1–14, 1989

2. David X. Wei and Steven H. Low, "A model for TCP model with burstiness effect," Submitted for publication, 2003.

3. Fernando Paganini, John C. Doyle, and Steven H. Low, "Scalable laws for stable network congestion control," in Proceedings of Conference on Decision and Control, December 2001, http://www.ee.ucla.edu/~paganini.

4. W. Feng and S. Vanichpun, "Enabling compatibility between TCP Reno and TCP Vegas," IEEE Symposium on Applications and the Internet (SAINT 2003), January 2003.

5. V. Jacobson, R. Braden, and D. Borman, "TCP extensions for high performance," RFC 1323, ftp://ftp.isi.edu/in-notes/rfc1323.txt, May 1992.

6. Lawrence S. Brakmo and Larry L. Peterson, "TCP Vegas: end-to-end congestion avoidance on a global Internet," IEEE Journal on Selected Areas in Communications, vol. 13, no. 8, pp. 1465–80, October 1995, http://cs.princeton.edu/nsg/papers/jsac-vegas.ps.

7. A. Kuzmanovic and E. Knightly, "TCP-LP: a distributed algorithm for low priority data transfer," in Proc. of IEEE Infocom, 2003

8. Y. Li, "Implementing highspeed tcp," URL:http://www.hep.ucl.ac.uk/~ytl/tcpip/hstcp/index.html.

9. L. Massoulie and J. Roberts, "Bandwidth sharing: objectives and algorithms," IEEE/ACM Transactions on Networking, vol. 10, no. 3, pp. 320–328, June 2002

10. R. Shorten, D. Leith, J. Foy, and R. Kilduff, "Analysis and design of congestion control in synchronised communication networks," in Proc. of 12th Yale Workshop on Adaptive and Learning Systems, May 2003, www.hamilton.ie/doug_leith.htm.

11. Shudong Jin, Liang Guo, Ibrahim Matta, and Azer Bestavros, "A spectrum of TCP-friendly window-based congestion control algorithms," IEEE/ACM Transactions on Networking, vol. 11, no. 3, June 2003.

12. R. Wang, M. Valla, M. Sanadidi, B. Ng, and M. Gerla, "Using adaptive rate estimation to provide enhanced and robust transport over heterogeneous networks," in Proc. of IEEE ICNP, 2002